Tess

and the MYSTERY Rock

by Melinda Bagby

Written by: Melinda Bagby
Illustrations: Jane Gross

Copyright © 2018 Gravitas Publications Inc.

Tess and the Mystery Rock

ISBN 978-1-941181-60-7

Published by Gravitas Publications Inc.
www.gravitaspublications.com
www.realscience4kids.com

Meet Tess!

Welcome to Real Science-4-Kids, where kids learn real facts about our world and explore how it works.

In this story, Tess is a curious girl who loves asking questions and exploring the world around her. Because she asks questions and explores her world, Tess is a scientist! You will discover how Tess does her exploration, the kinds of questions she asks, and the experiments she performs to prove or disprove her ideas, just like a grown-up scientist. You will also get a chance to do some experiments and learn about chemistry, physics, astronomy, biology, and geology, just like Tess.

This book has three different sections. The first section is the story of Tess, the mystery rock she finds, and the experiment she does to discover what kind of rock she has. The second sections are woven into the story and are called *Thought Stops*. These are sections where you will learn more about science and get to do experiments like Tess did. The third section at the end of the book is called *Science Facts*. These pages give you more information about rocks, bugs, what things are made of, and how they move.

What you will find by following Tess and her mystery rock is how easy it is to learn real science by asking questions, exploring facts, and playing with experiments.

Enjoy this reader and then check out our complete curriculum sets found at www.gravitaspublications.com.

Dr. Rebecca W. Keller, PhD
Author of the Real Science-4-Kids curriculum series

TESS and the MYSTERY Rock

My name is Tess.

I like to explore and
ask a lot of questions.
Mom says I'm like a
scientist. Scientists
ask lots of questions.

When I ask a lot of really good questions, Mom makes THE FACE. It's a special face she saves for when I do something fantastic...

...like paint!

... or wash the dog without even being asked. Mom makes THE FACE a lot.

Mom says scientists need to ask people lots of questions too.

4

She says Dad is a *wealth of knowledge* and I should ask him some questions. But it's hard to fit my questions in between the snores.

Scientists make observations. They look at things r-e-a-l-l-y closely and write down what they see. Then they think-think-think and come up with questions!

Mom gave me a notebook to write all my observations in. She's the best!

I take my notebook everywhere I go. Science is all around me, and I need to write down observations as soon as I see something super interesting.

observation:

you can see
more things
when you're
not staring at

a notebook.

Monday
observed a roly poly.
what do roly polys eat?
do roly polys live alone?

Where is your
family roly poly?
Ask Mom if i can
keep the roly poly.

Walking with a notebook
can be tricky at times!

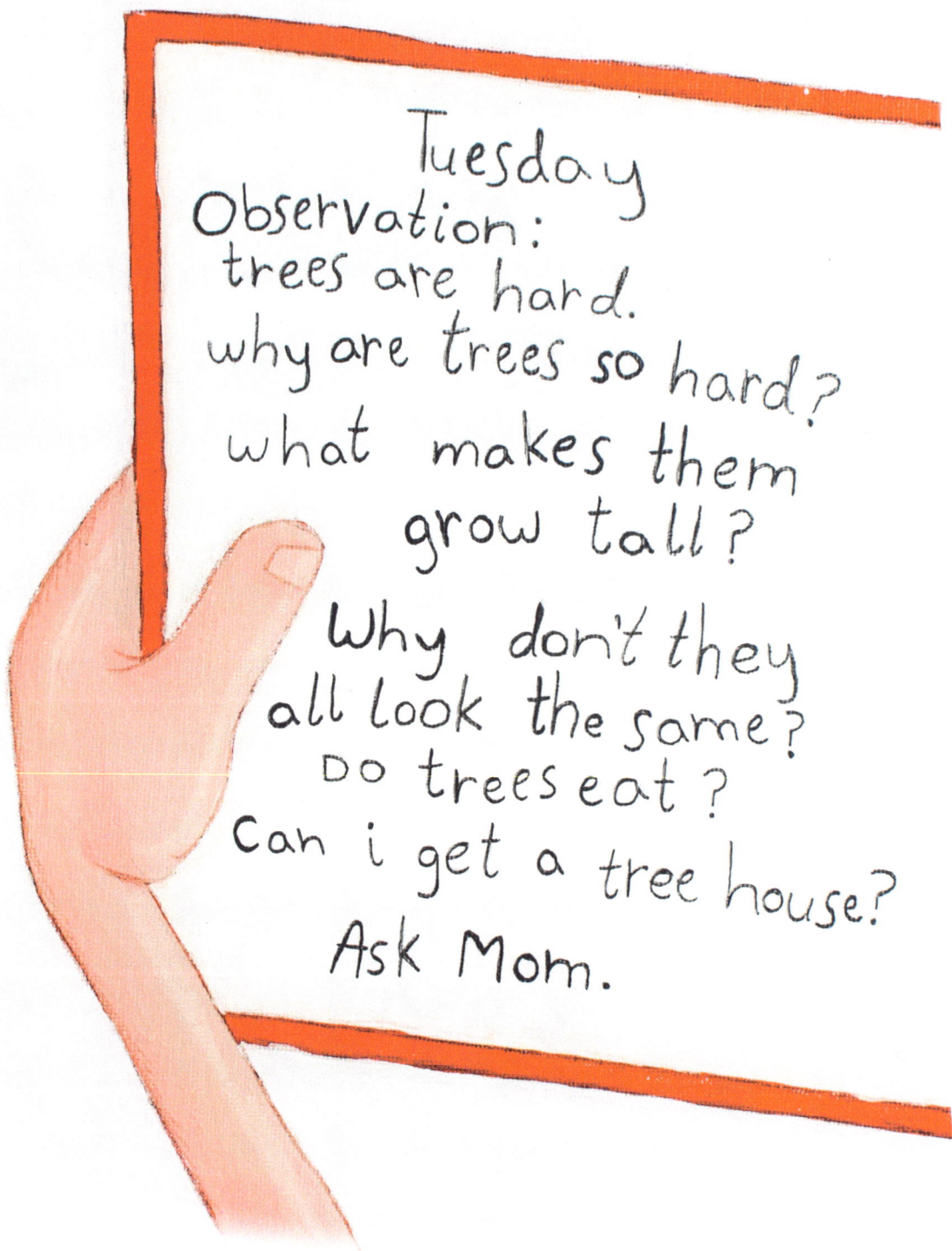

Tuesday

Observation:

trees are hard.

why are trees so hard?

what makes them grow tall?

Why don't they all look the same?

DO trees eat?

Can i get a tree house?

Ask Mom.

Walking to school
is a great time to
make observations.

Interesting things can come from anywhere!

Hmmmm...think-think-think....

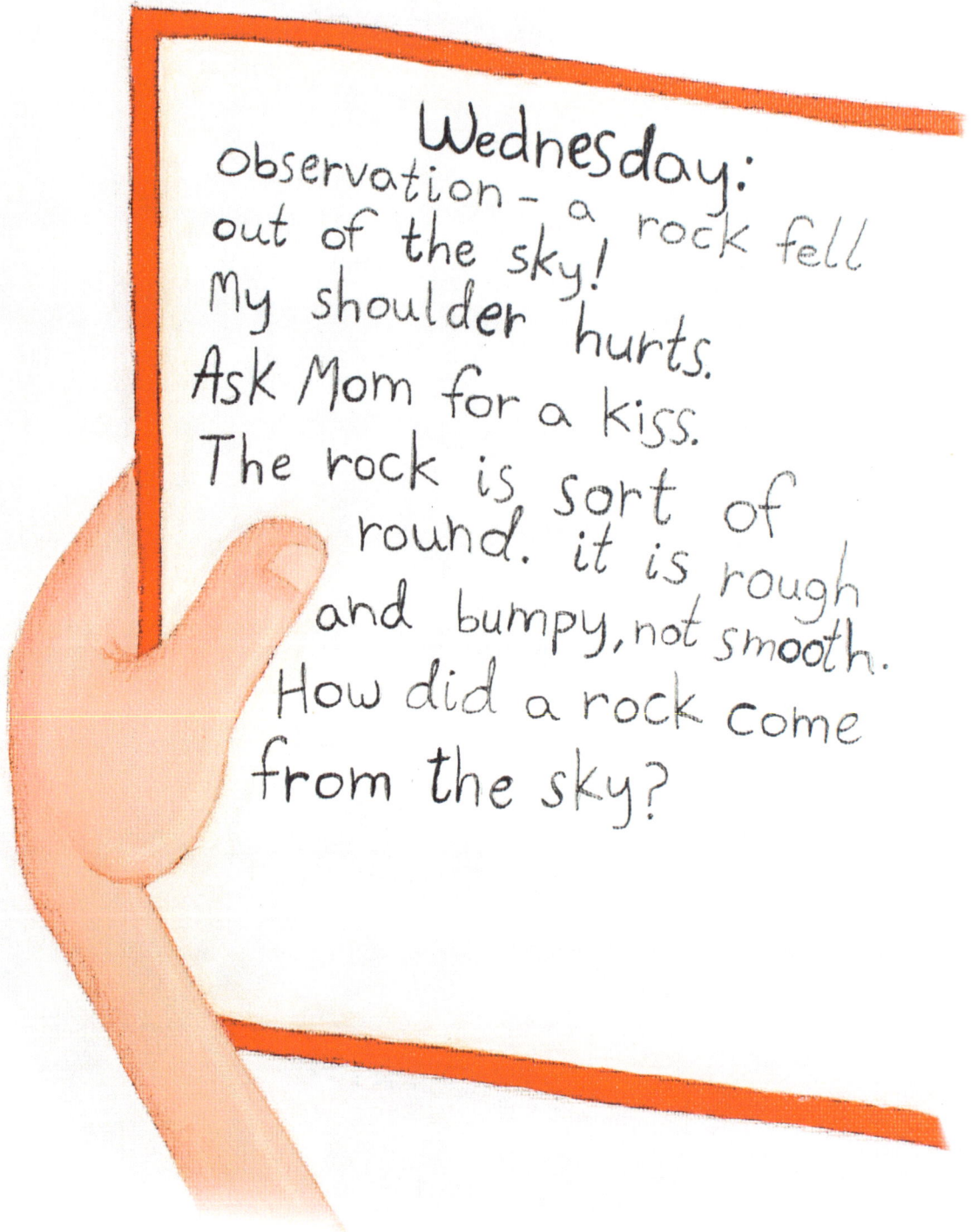

Wednesday:
observation - a rock fell
out of the sky!
My shoulder hurts.
Ask Mom for a kiss.
The rock is sort of
round. it is rough
and bumpy, not smooth.
How did a rock come
from the sky?

Dad says meteorites fall from the sky Meteorites are rocks. is this a meteorite?

Earth Rocks vs. Space Rocks

Earth Rocks...

- Are made by the Earth.

- Come in lots of different colors and shapes.

- Sedimentary rocks are made from bits of rock that have been pressed together. They often have layers. Some sedimentary rocks are round or oval and have crystals inside. Crystals are a type of rock-forming mineral and are often shiny with flat sides.

- Igneous rocks start out as melted rock deep within the Earth. They can be smooth or rough, and shiny or not shiny. They may contain crystals.

- Metamorphic rocks are made from other rocks that long ago sank deep into Earth and were melted. They may or may not have layers.

Space Rocks...

- Asteroids are rocks that are zooming through space.

- Comets are balls of ice and dirt moving through space.

- A shooting star is an asteroid or comet that has entered Earth's air and is burning up. A shooting star is also called a meteor.

- When a meteor hits the surface of Earth, the rock that remains is called a meteorite.

- Meteorites are heavy, often contain a lot of metal, and tend to attract magnets. May have dents in the surface that look like thumb prints. May look different from nearby rocks. May contain metal flakes or colorful, sphere-shaped bits of rock.

sedimentary

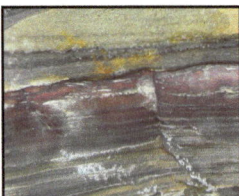

geode

igneous

metamorphic

meteorite

Photo credits: Geode and Meteorite, by H. Raab

Think About It

❶ How would you describe a meteorite to your friend?

_____

_____

_____

❷ How do you think a meteorite is different from other kinds of rock?

_____

_____

_____

❸ List the things that you think would help you tell whether or not a rock you find is a meteorite.

_____

_____

_____

❹ Do you think Tess found a meteorite? Why or why not?

_____

_____

_____

_____

...think-think-think! The rock is
round and bumpy. It fell from the sky.

It must be a meteorite!

I'm excited to show it to my teacher, Miss Anna.
My smile scrunches when I see Leo. He pulls my hair
a lot. I do NOT want him to see my meteorite.

"What's that?" Leo pries.

"It's an important scientific finding!" I curl my fingers extra tight around the meteorite.

Leo grabs my meteorite. The laws of motion seem to be working just fine, and Leo's applied force rips the meteorite out of my hand.

Observation: a two-handed grip would be stronger.

"Give that back, Leo!" I yell.

Leo looks unimpressed by the most fantastic scientific find of the year.

Miss Anna doesn't allow fighting, never ever. It's a classroom rule, a "get sent to the principal's office" kind of rule. She comes right over.

Speak FAST to gain the upper hand. This is the first rule of tattling. Also? Extend a pointer finger for effect.

"LEO took my BIG scientific finding that fell from the sky and I can't EVER EVER get a new one and it's a METEORITE!" I huff.

Miss Anna kindly retrieves the rock and
looks closely at it. Her face changes
from a hard face to a smile.
I like that face better.

She says, "I think this is
special, Tess, but I don't
think it's a meteorite.
Let's test my hypothesis."

"Meteorites contain metals and are solid rock on the inside.
Let's break open your rock and see what we find!

How To Be a Scientist

When scientists want to find out more about something, they follow a certain set of steps. These steps are called the scientific method.

❶ The first step is to make good observations. A good observation happens when you look at something in detail.

❷ The second step in the scientific method is to ask a question about what has been observed and then turn that question into a hypothesis. A hypothesis is a statement about something. For example, you might wonder, "Which is heavier, an elephant or a kangaroo?" Turning that question into a hypothesis, it becomes, "An elephant is heavier than a kangaroo."

❸ The third step in the scientific method is to do an experiment to prove or disprove the hypothesis. In the elephant-kangaroo example, the scientist would weigh the elephant and the kangaroo one at a time.

❹ The fourth step of the scientific method is collecting results. In the elephant-kangaroo example, to collect the results, the scientist would write down the weights of the two animals.

❺ The fifth and last step of the scientific method is to draw a conclusion. What did the information show? Was the elephant heavier than the kangaroo or was the kangaroo heavier than the elephant? The conclusion comes from the results of the experiment. Based on the results, the scientist draws a conclusion and shows that the hypothesis has been either proven or disproven.

Think About It!

Imagine you are with Tess and her friends. What do you think Tess and her friends will find inside the rock? Do you think it will be solid? Do you think the inside will look like the outside? Do you think it will be full of colored rocks? Do you think it will be full of water? Do you think it might have clay on the inside? Do you think something else might be inside the rock?

On the next page write down your hypothesis. Your hypothesis is your guess, or prediction, about what you think Tess and her friends will discover. Base your hypothesis on the information you have about Tess's rock and observations you make about it.

Draw what you think the inside of the rock will look like:

Observe It!

Hypothesis:

Tess and her friends will find _____
_____ inside the rock.

How would you observe the inside of the rock?

Again imagine you are with Tess and her friends. How many different things can you think of that you could do to try to open the rock? Which one do you think would work best?

❶ _____

_____

❷ _____

_____

❸ _____

_____

❹ _____

_____

❺ _____

_____

❻ _____

_____

Experiment!

❶ Think about what you have learned about the rock Tess found. What observations can you make about the rock?

_____

_____

_____

❷ Based on your ideas in the *Observe It* section, what things would you do to the rock to learn what is inside it?

_____

_____

_____

❸ What results do you think you might get by doing Step ❷?

_____

_____

_____

❹ Based on the expected results you listed in Step ❸, what conclusion would you draw about the rock Tess found? Would this prove or disprove your hypothesis?

_____

_____

_____

Miss Anna takes us outside and places my rock on the ground. Then she hits it with an even bigger rock.

My rock breaks into two pieces. It's filled with shiny white and purple crystals. It's not solid rock inside. Miss Anna's hypothesis is right! My rock isn't a meteorite, but it's beautiful!

Miss Anna calls it a geode and says the crystals are made of quartz. She says it's a hidden treasure, like me.

Wednesday #2:
Observation - some observations aren't what they seem to be. Even if you think you know something. Keep asking questions to make sure. it's O.K. to ask questions. it's O.K. to be wrong.

i like Miss Anna.
She makes me
feel good.

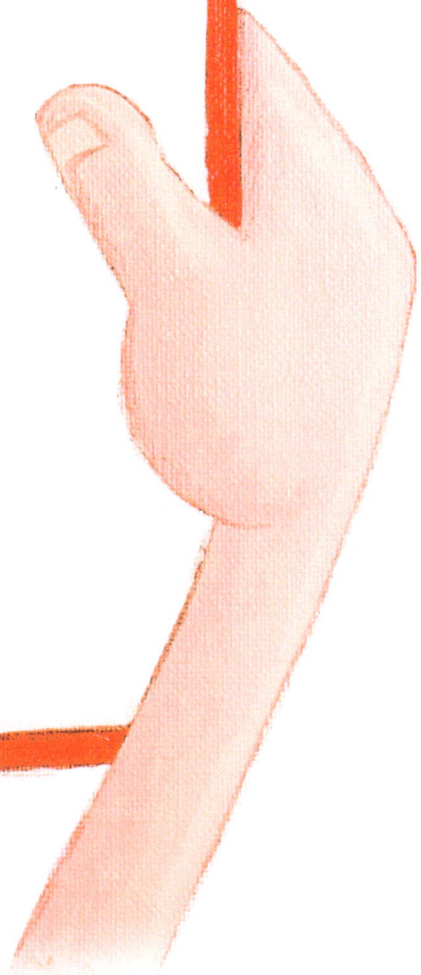

What did you discover?

❶ What kind of rock did Tess find? _____

❷ Was your hypothesis proven or disproven? Why?

_____

_____

❷ How would you compare the rock Tess found to a meteorite?

Geode	Meteorite
_____	_____
_____	_____
_____	_____
_____	_____
_____	_____

❹ Draw a picture of the inside and outside of the rock Tess found.

Science Facts
To Think About!

Discover Roly Polys

If you have ever seen a small bug run across the ground and then roll into a ball, you might have spotted a roly poly.

Roly polys go by different names. Depending on where you live, they can be called butchy-boys, doodle bugs, or pill bugs.

A roly poly is a type of crustacean [crust-AY-shun] that lives on land. A crustacean is an animal that has a shell to protect it, a body divided into segments, and two pairs of antennae. Most crustaceans, like crabs and lobsters, live in rivers, lakes, or oceans, but a few, like the roly poly, live on land.

When a roly poly feels threatened, it will curl into a tight ball to protect itself. They are gentle animals and won't bite or sting. They just like to go about their business, digging in the dirt and looking for moss, bark, or decaying plants to eat. You are most likely to find them in moist places. Look under a rotting log or beneath a pile of leaves.

Photo credits: 1. Franco Folini, 2 & 3. Katya from Moscow, Russia

Meteorites

If you have ever looked up into the night sky and watched a shooting star, you might have been observing a meteor passing through the atmosphere. A meteor starts out as an asteroid, which is a rock traveling through space. If an asteroid enters Earth's air, you can see it as it burns up and becomes a meteor, or shooting star. Sometimes small chunks of the meteor survive the journey and hit the ground. When that happens, the rock on the ground is called a meteorite.

Meteorites are made of iron or rock-forming minerals or a mixture of both. Meteorites are the rarest type of rock found on Earth! Tess discovered that it can be hard to tell if a rock is a meteorite or just a regular rock, but sometimes a simple test can help you find out.

Meteorites are usually heavy for their size, may look like the outside has been melted, and are often dark in color. They have an irregular shape, may have thumb print-like pits, and are solid inside. Metal flakes or colorful, sphere-shaped bits of rock can be present in some meteorites. One of the best ways to test if a rock might be a meteorite is to see if it is magnetic!

Photo credits: 1. Meteor, by Navicore;
2 & 3. Meteorites, by H. Raab

Geodes

When you pick up a rock, it can be fun to think about what might be inside. Most of time when you break a rock open, it looks the same on the inside as on the outside. But sometimes you get lucky, and when you crack the rock open, you find it is hollow and lined with crystals! This type of rock is called a geode.

Geodes can be ball-shaped or oval and look very ordinary on the outside. A geode can have its beginning in an area of volcanic activity or in a deposit of mud that is later turned to rock. Often a bubble of gas is trapped as the rock is formed, resulting in a hollow space in the center. Later, groundwater seeps through the hardened surface of the rock into the hollow interior. Groundwater carries minerals, and over very long periods of time these minerals can form crystals in the hollow center. Geodes come in many different colors, depending on which minerals are in the groundwater. Geodes are found all over the world.

Photo credits: 1. Geode–outside, by I, Manfred Heyde; 2. Amethyst quartz inside geode, by Nkansah Rexford; 3. Amethyst quartz crystals, by Didier Descouens; 4. Geode, by Rob Lavinsky, iRocks.com, CC-BY-SA-3.0; 5. Geode, by Kora27

Making Geodes

Have you ever wondered what everything is made of? You and all the things around you are made of very, very tiny building blocks called atoms. Atoms hook together to make molecules.

Molecules can be very simple and made of as few as two atoms, or they can be very complicated and made of thousands of atoms.

Try this little experiment. Put some water in a glass. Put some salt on a spoon and observe it. You can see that the salt is in small pieces called crystals. Now stir the salt into the water.

What happens to the salt? Although the salt seems to disappear, if you taste the water, it tastes salty. So the salt is still there, but where is it? The salt molecules in the salt crystals have broken away from each other and moved in between the water molecules. We say the salt has dissolved in the water. Because individual molecules are too small for you to see with your unaided eyes, you can't see either the salt molecules or the water molecules.

In a similar way, rock-forming atoms and molecules can dissolve in water. When water carrying the right kinds of atoms and molecules enters a hollow part of a rock, the atoms and molecules are carried in with the water. When the water leaves, the rock-forming atoms and molecules that are left behind form crystals.

sodium   chlorine

A salt molecule has one sodium atom (Na) and one chlorine atom (Cl).

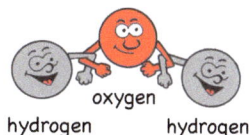

hydrogen   oxygen   hydrogen

A water molecule has one oxygen atom (O) and two hydrogen atoms (H).

Silicate is a rock-forming molecule that has 4 oxygen atoms (red) and one silicon atom (green).

Newton's Laws of Motion

What happens when you throw a ball in the air? Does it go up, up, up and out into space or does it come back down? Try it!

When you throw a ball up in the air, it always comes back down. In fact, without help from something like a rocket engine, you can't throw a ball hard enough to get it to go out into space. A ball thrown in the air follows Newton's Laws of Motion.

Isaac Newton was a physicist and mathematician born in England in 1643. He thought a lot about how things move, why they move, and what keeps them moving. All of this thinking led him to develop a set of mathematical "laws." These laws explain the motion of objects and are called Newton's Laws of Motion.

In the story, when the rock flies off the truck and strikes Tess, it is obeying Newton's First Law of Motion that states: "An object at rest (not moving) will stay at rest unless an external force is applied to it."

If we think about the rock in the truck, we can visualize the rock sitting happily on the bed of the truck until the truck starts moving. The vibrations in the truck jiggle the rock until it falls off, bouncing off the ground and hitting Tess on the shoulder. The moving truck and the vibrations are "external forces" that eventually bounce the rock off the truck. If the truck didn't move or vibrate, there would be no external forces and the rock would stay at rest on the truck.

asteroid [AS-ter-oyd] • a rock that is moving through space

atom [A-tum] • a basic building block of a substance

comet [KAH-met] • a ball of ice and dirt that is moving through space

conclusion [kun-CLUE-zhun] • the last step in the scientific method; the summary of the results or outcome of an experiment

crustacean [crust-AY-shun] • an animal that has a shell to protect it, a body divided into segments, and two pairs of antennae; examples are roly polys, lobsters, and crabs

crystal [KRIS-tuhl] • a type of mineral that has an organized internal structure of atoms; can have smooth, flat sides; can be shiny

dissolve [di-ZAHLV] • when the molecules in a solid break apart from each other and go in between the molecules of a liquidw

experiment [iks-PER-uh-ment] • a step-by-step series of actions used to prove or disprove a hypothesis or idea

force • the power that changes the position, shape, or speed of an object

geode [JEE-ode] • a rock that has a hollow interior that is lined with crystals

hypothesis [hi-PAH-thuh-sis] • a statement about something that has been observed; a prediction or guess about what the results of an experiment will be

igneous rock [IG-nee-us rock] • rock that has its beginning as melted rock deep within the Earth; igneous rocks can be smooth or rough, shiny or not shiny, and may contain crystals

law of motion • a scientific statement that describes how objects move and what makes them move

metamorphic rock [met-uh-MORE-fik rock] • a type of rock that is made from other rocks that long ago sank deep into Earth and were melted; a metamorphic rock may or may not have layers

meteor [MEE-tee-or] • an asteroid that has entered Earth's air and is burning up; also called a shooting star

meteorite [MEE-tee-or-ite] • a rock that results from a meteor reaching the surface of Earth

mineral [MIN-rul] • a naturally occurring, solid material that has an organized internal structure of atoms; found in rocks

molecule {MAH-luh-kyool] • two or more atoms joined together

observation [ob-sur-VAY-shun] • the act of looking at something carefully and in detail

quartz [kwortz] • a common mineral

rock-forming mineral [MIN-rul] • a mineral commonly found in rocks

scientific method • a series of steps used by scientists to make discoveries

sedimentary rock [sed-uh-MENT-uh-ree rock] • a type of rock that is made from bits of rock that have been pressed together

shooting star • a meteor

Real Science is FUN!